DOLLS A TO Z

By

Mary Bertha Brown

ISBN: 1-4107-5189-9 (e-book)
ISBN: 1-4107-5188-0 (Paperback)

Library of Congress Control Number: 2003094397

This book is printed on acid free paper.

Printed in the United States of America
Bloomington, IN

1st Books - rev. 06/16/03

Dedication

I dedicate my book "Dolls A to Z" to my deceased mother, Rosa Cunningham Brown. My mother loved and appreciated dolls. She gave me my first doll, which inspired me to begin a collection of twenty-six dolls.

Forward

I was inspired to write "Dolls A to Z" because of my mother giving me that very first doll to start my collection. My mother is deceased but the collection has grown from one to twenty-six dolls. Twenty-four of the dolls are Black. Two of the dolls are White and they were given to me by two of my mother's best white friends.

Each of the twenty-six dolls has a name beginning with a letter of the alphabet. One of the dolls is certainly named after my deceased mother, Rosa. The two White dolls are named after my mother's best friends, Bell and Mary. I gave the other twenty-four dolls names to match the remaining letters of the alphabet.

I wrote this book as a result of my determination to collect dolls, and for the fact that dolls are valuable and meaningful to many people throughout the universe.

In reading this book people will appreciate how precious and valuable dolls can be in a collection base on custom, history, values and traditions.

Mary Bertha Brown

Introduction

"Dolls A to Z" is written for anyone who likes to collect dolls. The twenty-six dolls are given names of the alphabet from A to Z. Twenty-four of the dolls are Black and two of the dolls are White.

One of the twenty-six dolls is named after my deceased mother, Rosa. The two White dolls are named after my mother's two best friends, Bell and Mary.

Each of the dolls has a description about them from their clothes to their physical appearances.

This book about dolls allows people and individuals from many phases of the universe to know how valuable and meaningful dolls can be.

My conclusion is that the readers and people in general will find out how precious and special dolls can be in a collection base on custom, history, values, and traditions.

Mary Bertha Brown

A is for Addie.

Addie

Addie is a doll with a flat brown face, small brown sauce eyes, button nose, small half moon mouth, brown curly hair, short brown arms, and short brown legs.

She is wearing an "ABC" cotton dress with a half moon collar in the colors red, white, and blue. She is wearing black high top shoes with white shoe laces. Her dress has apples over it. The front hem of Addie's dress has the number 1, 2, 3, written in the color blue.

Addie is an adorable doll for any little girl or for any pre-school or Kindergarten classes.

Mary Bertha Brown

B is for Bell.

Bell

Bell is a white doll made out of cloth, stuff with cotton in the inside. She is name in the honor of my mother's decease friend "Bell." Bell has a wide flat face, big blue saucer eyes, small flat nose, small pink lips, pink rosey cheeks, and short curly hair.

She is wearing a long white cotton dress with mixure of the color blue, with ruffles and eye in bed pink ribbon, and white matching bonnet with a pink ribbon. She is carrying a pink raddle in her left hand.

She is a soft beautiful doll for any little girl to hold or just curled up with or just lay on her bed.

Mary Bertha Brown

C is for Catherine.

Catherine

Catherine is a Black bridal doll with a wide brown flat face, box shape brown nose, wide red lips, and bright brown oval eyes. She has long browish curly hair with a Peter Pan bangs.

She is wearing a long white floor length satin Wedding gown, with shoulder length sleeves, a big crystal like bow hanging down, flat white shoes with T-straps, and matching white nylon socks. She is wearing a white throw back vale made out of nylon.

Catherine is an eye catching doll for any preschool child or any adult. She could be on display in anyone's home, museum, or a Bread and Breaksfast Inn.

Mary Bertha Brown

D is for Donna.

Donna

Donna is a Black doll with a brown oval like shape face, big brown saucer eyes, small box shape nose, small pin like lips. She has long black curly hair and a beautiful curly bangs.

She is wearing a pink sweat shirt with V-neck collars, a little boy wearing an over all on front, with long denium blue pants. She has on matching white T-strap shoes with white nylon socks. She is wearing two bright red bows in her hair.

Donna is an adorable doll for any little girl, preschool, Kindergarten class rooms, or just to pit in any collection.

Mary Bertha Brown

E is for Eve.

Eve

Eve is a small Black baby doll with a flat small oval like shape face, pointed brown nose, oval shape brown lips, fat baby cheeks, brown bubble eyes, and baby stringy brown hair.

Eve is wearing a sleeveless nylon dress with Spring flowers on it and clown collar. She is wearing a colorful matching short vale. She isn't wearing shoes.

She is a cute baby like doll for any little girl to curdle up with during day or night. She is adorable for any preschool or Kindergarten classrooms Housekeeping centers or anyone's doll collector's item.

Mary Bertha Brown

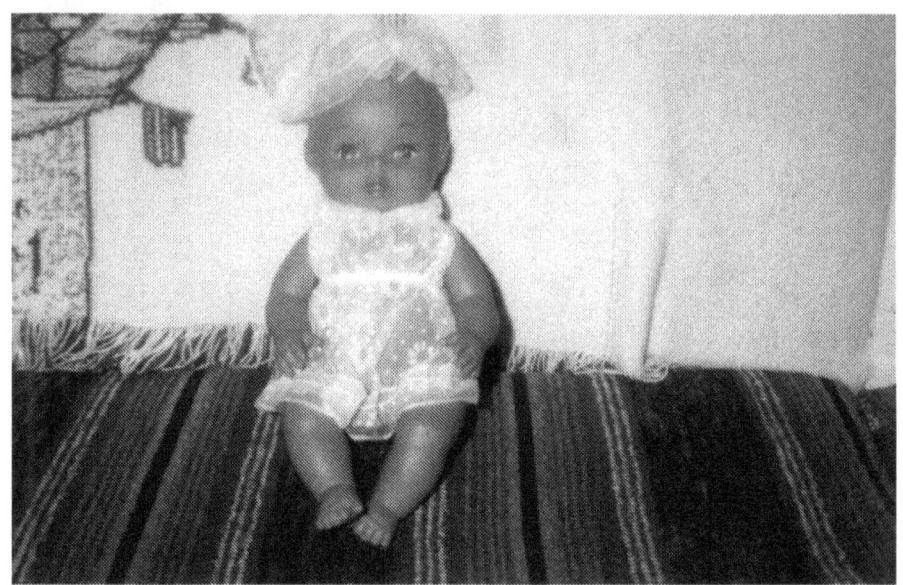

F is for Francis.

Francis

Francis is a Black doll with a small circular brown face, big brown circle eyes, small box like nose, small red pear shape lips, small rosey cheeks, and short black curly hair with a cut bangs.

Francis is wearing a sleeveless cotton dress with no collar, top of the dress is white, bottom of her dress are blue and white check with a red threat hem. She is wearing egg shell white matching boots.

She is a charming doll for any little school girl to play with at all times. She is eye catching to put in any doll's collection.

Mary Bertha Brown

G is for Georgia.

Georgia

Georgia is a Black doll twin to her sister Donna. She has a brown oval like shape face, big brown saucer eyes, small box shape nose, small pin like lips. She has long black curly hair, and a beautiful curly bangs.

She is wearing a pink cotton polka-dot dress made out of cotton, no collar, puff short sleeves, and a lace band in front, white T-strap matching shoes with white nylon socks, and red satin ribbon bows.

Georgia is a beautiful doll for any little girl, preschool, kindergarten class rooms, put on any display or for any doll's collection.

Mary Bertha Brown

H is for Hope.

Hope

Hope is a Black baby doll with a big circular shape brown face, precious realistic brown eyes, corn shape box brown nose, brown pear shape lips, short straight brown hair, and bow legs.

She is wearing a pink orylon crawl in suit with lace ruffles, bit white bow with flowers attach to it, a white blouse with red flowers, matching pink head band with a big white bow that has red flowers on it. She is wearing no shoes nor booties.

Hope is designed for holding, hugging and loving, baby feels soft as a real bundle of joy, with powder fresh scent and realistic brown eyes she'll be any little girl most precious love. She is adorable for any preschool or Kindergarten classrooms, for adults, or any collector's corner for display.

Mary Bertha Brown

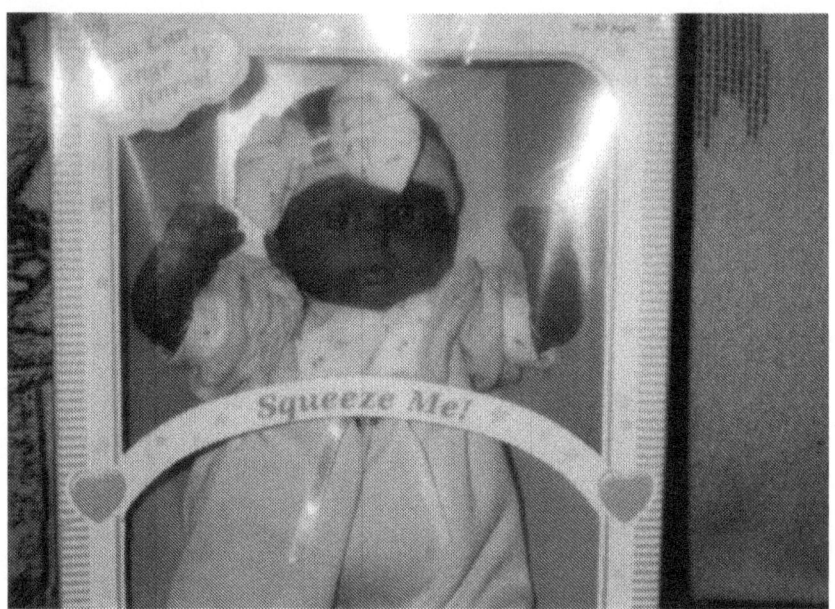

I is for India

India

India is a Black rag doll with a wide flat pecan brownish face, big brown saucer eyes, small circular nose, red half moon lips, long black mop like hair with bangs.

She is wearing a purple cotton print dress with yellow, blue, and purple flowers with match bonnet and cloth matching shoes. She is wearing matching ribbons in her hair in the color pink and matching pink ribbon in front of her dress.

India is an eye catching doll for any little girl to cuddle and hold for teenage girls or any adult lady. She is ideal for any preschool or Kindergarten classes. She is an adorable doll for any collection or just to be on display in any children's museum.

Mary Bertha Brown

J is for Jennifer.

Jennifer

Jennifer is a Black doll with big brown circular shape face, big brown oval eyes, a triangular box shape nose, wide pear shape red lips, long black hair with pony tails hanging down, and a big but short bangs.

She is wearing a blue and pink flowers dress, with no collar, made out of cotton, puff white eyelet sleeves, with matching baby blue ribbons hanging down, like matching baby blue ribbons in her hair. She is wearing white T-strap shoes with white nylon socks.

She is a prescious doll for any little girl to hold, curdle up with, to play house and have an enjoyable tea party. She is adornable for anyone's dolls collection.

Mary Bertha Brown

K is for Katrina.

Katrina

Katrina is a Black rag doll with a flat pecan brownish face, big brown button eyes, wide flat box nose, brown half moon lips, long brownish mop like hair with bangs.

She is wearing a cotton print mangle dress with different colors, with puff long sleeves, pink band around it, with matching mangle hat of different colors. She is wearing long lavender pants with small white poka-dot and matching pink botties. She has matching pink ribbons in her hair.

Katrina is an eye catching doll for any little girl, teenage girls or any adult ladys. She is adorable for any preschool or Kindergarten classes. She is a treasable doll for any collection or just to be on display in any children's museum.

Mary Bertha Brown

L is for Lucy.

Lucy

Lucy is a Black doll with an oval brown face, glassy brown eyes, small brown triangular nose, puff brown cheeks, small brown pin lips, and long brown curly hair with a big brown circular bangs.

She is wearing an egg shell white long satin dress with lace ruffles, lace collar and lace down the front of her dress. She is wearing a short red velvetine jacket with matching red velvetine tam, matching egg shell white lace stockings, and matching egg shell white boots like shoe with T-straps.

Lucy is an adorable doll for any little girl, girls of any age, ladies, and elderly ladies. She is eye catching to be place on display in any museum or curio. She is ideal for any collector's corner or suitable for any little girl's doll house or bedroom.

Mary Bertha Brown

M is for Mary.

Mary

Mary is a white doll name in the honor of my mother's best friend. She has a big broad round face, big saucer blue eyes that open and close, small button like triangular nose, small round orange lips that opens, pug orange cheeks, blonde curly hair tie in two pony tails tie up with a big curly bangs.

She is wearing a short yellow cotton dress with white, puff sleeves, high lace collar, white lace around the hem, pink flowers in front of the dress, white ribbons matching in her hair, white matching nylon socks, and she has a white plastic bottle by her side.

Mary is an adorable doll for any little girl to curdle and hold to have a tea party with at times. She is precious for any preschool or Kindergarten class. She is ideal for any collection or any dramatic play area. She could be on display in any children's museum.

Mary Bertha Brown

N is for Nova

Nova

Nova is a Black doll laying in a straw bassinet with straps. She has a big Brown circle face, big Brown saucer eyes that open and close, small button like nose, small orange lips, puff orange cheeks, short black curly hair.

She is wearing a sleeveless cotton dress with black and white straps at the top and blue, black, white, and pink plaid, she isn't wearing any shoes.

Nova is an eye catching doll for any preschool or Kindergarten classrooms dramatic play area or Housekeeping centers. She is adorable for anyone's doll collection, curio, or just for any little girl's doll house or just to curdle up with at day or night time.

Mary Bertha Brown

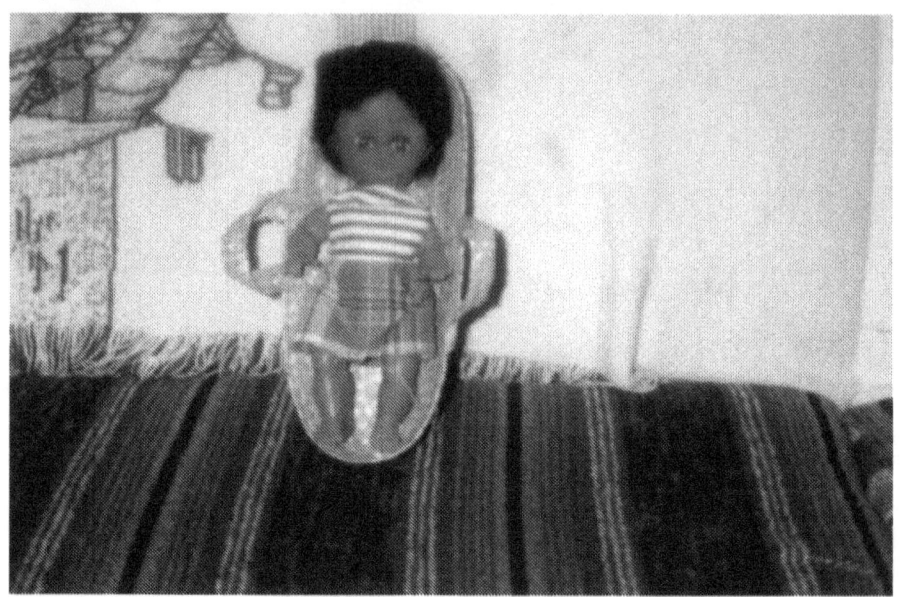

O is for Oliva.

Oliva

Oliva is a Black doll with a small Black triangle face, small Black glassy eyes, small pointed triangle nose, small burgundy lips, long straight Black hair with a big curly bangs.

She is wearing a short pink polk-a-dot blouse with short puff sleeves, deep pink shorts with a short waist line with a big white button in front. She isn't wearing any shoes.

Oliva is a jazzy doll with a fancy style, and smile for a girl of any age. She is adornable for any collection and for any children's museum to be place on display. She will be amazing to be on display in any lady's curio.

Mary Bertha Brown

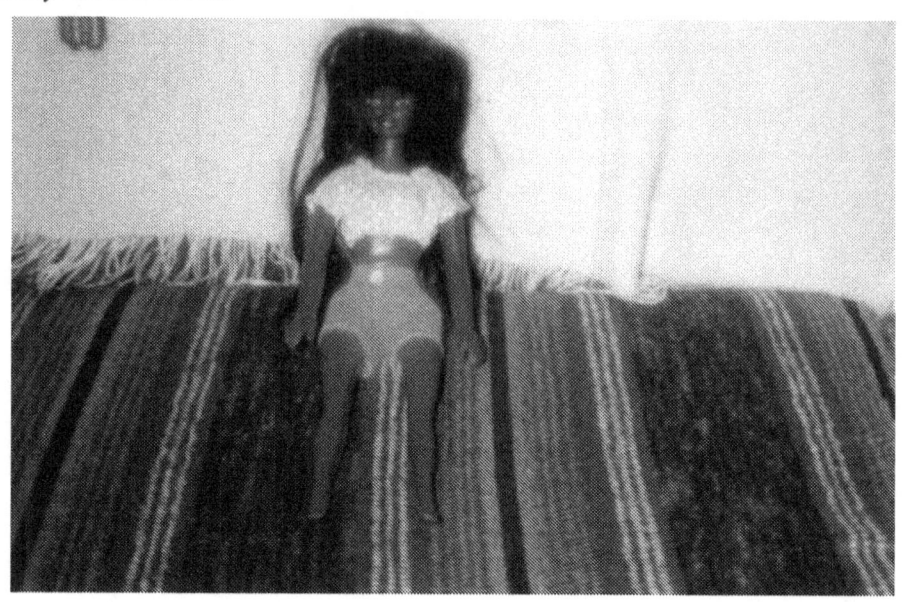

P is for Patrice

Patrice

Patrice is a Black doll with a big round brown face, big saucer brown eyes, small pin button nose, wide purple moonshape lips, puff purple like cheeks, long black straight hair with wide curly bangs.

She is wearing a pink cotton long sleeves blouse with matching short sleeves valore pink top, long cotton off-white pants with blue strips down the sides, white patten T-strap shoes with pink matching nylon socks, and yellow satin ribbon in her hair.

Patrice is a beautiful teenage looking doll for girls of any age to have around a home, playrooms, classrooms, museums or to add the finishing collection to anyones curio. She is ideal to be use for any "Show and Tell."

Mary Bertha Brown

Q is for Queen.

Queen

Queen is a Black doll with a small brown triangle face, small brown oval eyes, small triangle nose, small moon shape burgundy lips, long black hair with a wavelength bangs.

She is wearing a sleeveless green and white shoulder dress made out of fleece material with a wide silver belt with matching green and white fleece hat with a white tassel. She is wearing white matching pumps with white matching bows.

Queen is a doll look like a Las Vegas show girl that is ideal for any girl of age or any grown up lady. She is adorable for any curios, corner or shelf. She is ideal for any dolls collection or for any children's museum.

Mary Bertha Brown

R is for Rosa.

Rosa

Rosa is a Black doll name in the honor of my decease mother. She has a big broad brown round face, big saucer Brown eyes that open and close, small button like triangular nose, small round Brown lips that opens, pug Brown cheeks, Black curly hair tie in two pony tails tie up with a big curly bangs.

She is wearing a short white cotton dress with White, puff sleeves, high lace collar, White lace around the hem, pink flowers in front of the dress, Red ribbons matching in her hair, White matching nylon socks, and she has a white plastic bottle in front of her.

Rosa is an adorable doll for any little girl to curdle and hold, to have a tea party with at times. She is precious for any preschool or Kindergarten classes. She is ideal for anyone's curio's shelf or any children's museum.

Mary Bertha Brown

S is for Sabrina.

Sabrina

Sabrina is a Black cabbage patch kids doll with a wide flat oval Brown face, small Brown circle glassy eyes, small Brown triangle nose, small Brown half moon lips, small Brown puff cheeks, and long stringy Black hair with a short cut bangs.

She is wearing a cotton print dress in the colors Red, White, and Blue looks like a handerchief with a big blue and white bow in front, with long puff sleeves. She is wearing Black patent T-strap shoes with matching white nylon socks. She has matching Red satin ribbons in her hair. She has a yellow brush, yellow mirror, red hair drier, and a deep pink bottle of hair spray all beside her.

Sabrina is a dream doll for anyone's collection of dolls for any little girl's toy chest or doll house. She is adorable for any preschool or Kindergarten classrooms. She is ideal for any children's museum.

Mary Bertha Brown

T is for Telma

Telma

Telma is a Black rag doll with an oval brown flat face, a circular shape mouth, small brown saucer eyes with an orange glow to them, plump nose, short fat cheeks, and long black curly hair hanging down in matching pony tails.

She is wearing a red cotton patch print dress with a penafold design with a pink matching crawl in suit underneath. She is wearing a pink and red cotton patch print bonnet on her head with a red ribbon on one of her pony tail. She is wearing matching red cotton patch print shoes.

Telma is an adorable doll for any little girl to curdle up with, to hold, or have a tea party with. She is ideal for any preschool or Kindergarten classes housekeeping or Dramatic play centers. She is eye catching for any children's museum, curio, or for any dolls collection.

Mary Bertha Brown

U is for Urica

Urica

Urica is a Black doll with an oval like flat face, big brown saucer eyes, medium size broad nose, medium size half moon red lips, puff cheeks, very short eye burls, and long black hair with matching braids with matching blue ribbons.

She is wearing a blue and yellow penafold dress that have palm trees and suns in it. Her dress has long yellow puff sleeves and a yellow clown collar. She is wearing a pink crawl in suit underneath. She is wearing matching cotton print shoes with palm trees and suns in it like her dress.

Urica is an adorable doll for any little girl to curdle up with, to hold, or have a tea party. She is ideal for any preschool, Kindergarten, or child care classes housekeeping or Dramatic center's. She is a dream doll for anyone's collection of dolls, for any little girl's toy chest or doll house. She is ideal for any children's museum, curios, or ideal to be use for any "Show and Tell."

Mary Bertha Brown

V is for Victoria

Victoria

Victoria is a Black doll with a triangle like face, small dark brown eyes, small pin shape nose, small half moon wine color lips, small cheeks, very short black eye burls, long black hair comb back with a baby blue ribbon tie around her hair.

She is wearing a short orange swim suit with matching top in the colors blue, yellow, green, plum, dark pink, white with a light green ribbon in front and a half moon collar in the color orange. She is bear feet, but she has white T-strap high hell pump shoes. She has a yellow towel, blue and white circle like flot, blue earphones with a matching blue CD disc radio, yellow brush and comb laying out around her on the beach.

Victoria looks like a model. She is an adorable doll for little girls and preteenage girls. She is very eye catching to play with at anytime. She is a dream doll for anyone's collection, children's museum, play room, or for "Show and Tell."

Mary Bertha Brown

W is for Wendy

Wendy

Wendy is a Black doll with a round circular face, medium size brown eyes, a medium box nose, medium half moon wine color lips, puff brown cheeks, long brown eye burls, short brown hair with a bangs cut short.

She is wearing a long white crawl in suit with matching white hat. Her white crawl in suit has long white sleeves, and long white puff legs. She is bear feet. Her crawl in suit has three white flowers that looks like daisies.

Wendy is an adorable baby looking doll for any little girl to curdle up with, to hold, to play house with any time of the day or night. She is ideal for any preschool, Kindergarten or child care classes housekeeping or Dramatic centers. She is a dream baby doll for anyones collection of dolls, for any little girl's toy chest or doll house. She is ideal for any children's museum, curios, or ideal to be use for any "Show and Tell."

Mary Bertha Brown

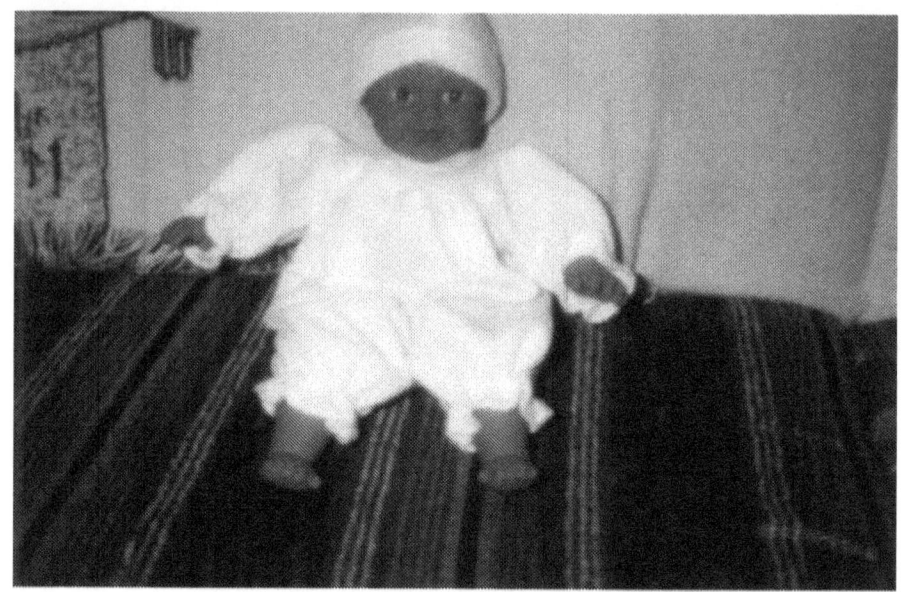

X is for Xavier

Xavier

Xavier is a tall Black doll with a broad brown oval like face, big brown saucer eyes, small light brown eye burls, medium burgundy pear shape size lips, broad box size brown nose, big brown half moon ears, long black Shirley Temple curly hair with a broad cut bangs.

She is wearing a long length burgundy velveteen dress that has long sleeves with lace, a high eye shell white satin collar with a lace flower in front, the dress has three ruffles with eye shell white lace, and pleated eye shell white satin below the ruffles in front of the dress. She is wearing long legged eye shell white panies with lace at the botton, long legged eye shell white nylon stockings, matching white high top shoes with white nylon shoe laces. She is wearing a small white feather hair piece with a pink flower that has a green satin ribbon and pink ribbon.

Xavier is an adorable doll for anyone's collection of dolls, curio's shelf, for any little or teenagers girl to curdle up with, play house or school anytime of the day or night. She is ideal for any children's museum, Bread and Breakfast Inn, or ideal for any lady, or for any "Show and Tell."

Y is for Yolanda

Yolanda

Yolanda is a small black baby doll with a round circular brown face, medium size brown eyes, a medium size brown nose, no eye burls, very short brown baby like hair, puff brown like cheeks, big brown circular ears, and big brown puff lips.

She is wearing a white over haul looking pants with straps, that has small dark pink hearts, with a matching light pink checked long sleeves blouse with elastic in both sleeves of the blouse that has matching white Peter Pan like collars. She is wearing a white head band with a light pink checked bow around her head. She is wearing matching light pink checked cloth shoes.

Yolanda is an adorable baby doll for any little girl to curdle up with, to hold, to play house with anytime. She is ideal for any preschool, Kindergarten, or child care classes. Housekeeping or Dramatic Art Centers. She is a baby doll for anyones collection of dolls, for any little girl's toy chest or doll house. She is deal for any children's museum, curios, or ideal to be use for any "Show and Tell."

Mary Bertha Brown

Z is for Zena

Zena

Zena is a black doll with a medium size round circular face, small black glassy eyes, small black eye burls, a small brown pointed triangle nose, small pink lips, small brown circular ears, long black crystal hair tie in a ponytail with a short Las Vegas show girl bangs hair cut.

She is wearing a short sleeves dress with pink, purple, and blue flowers in it. The dress has a bright orange band at the top. She is wearing dark pink stick in pumps shoes. She has a long bright pink hair comb to comb her hair.

Zena looks like a Hollywood or New York model doll. She is an adorable doll for any little girl to play house, have a tea party, or just to curdle up with anytime of the day or night. She is an ideal doll for anyone's collection, curio's shelf, any children's museum, a Bread and Breakfast Inn, a playroom, or for any "Show and Tell."

Mary Bertha Brown

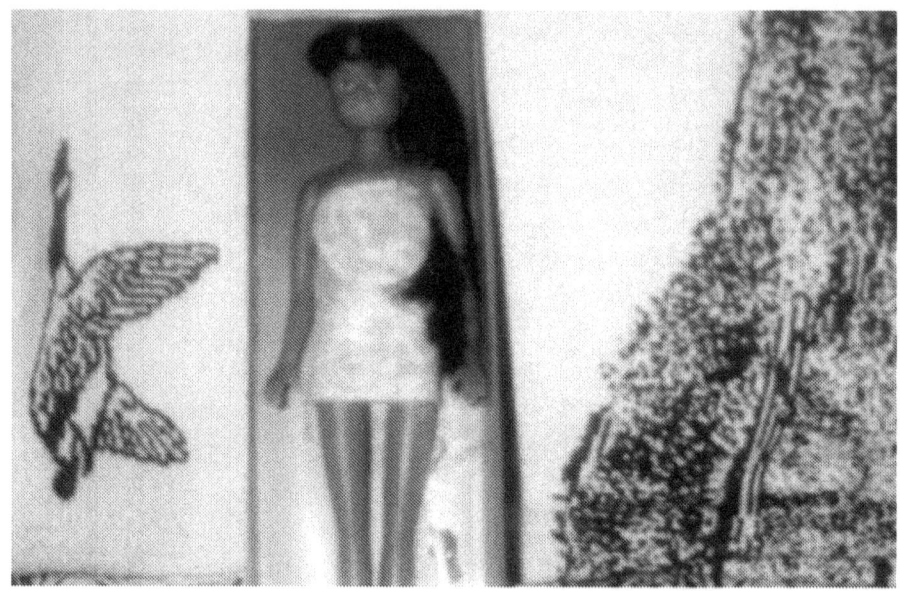

About the Author:

Mary Bertha Brown is a native of Georgetown, South Carolina. She is presently living in Carvers Bay Community of Georgetown County. Mary is a former Teacher, Educator, Early Childhood Specialist, and Day Care Owner and Director. She has worked with the YMCA and Pepper Geddings Recreational Center in Myrtle Beach, South Carolina.

Mary was a classroom teacher for thirteen years in the public and private schools. She won many awards during her years as a classroom teacher.

Mary Bertha Brown is a graduate of St. Augustine's College with a Bachelor of Arts Degree in Social Studies. She also has a Masters of Education in Early Childhood Education from the University of South Carolina.

Mary is presently working as a Child Care Specialist at a Group Home for Boys in Conway, South Carolina.

Dolls A to Z is her first book.